EARLY HOME COMPUTERS

Kevin Murrell

Published in Great Britain in 2013 by Shire Publications Ltd, Midland House, West Way, Botley, Oxford OX2 0PH, United Kingdom.

43-01 21st Street, Suite 220B, Long Island City, NY 11101, USA.

E-mail: shire@shirebooks.co.uk   www.shirebooks.co.uk

© 2013 Kevin Murrell.

All rights reserved. Apart from any fair dealing for the purpose of private study, research, criticism or review, as permitted under the Copyright, Designs and Patents Act, 1988, no part of this publication may be reproduced, stored in a retrieval system, or transmitted in any form or by any means, electronic, electrical, chemical, mechanical, optical, photocopying, recording or otherwise, without the prior written permission of the copyright owner. Enquiries should be addressed to the Publishers.

Every attempt has been made by the Publishers to secure the appropriate permissions for materials reproduced in this book. If there has been any oversight we will be happy to rectify the situation and a written submission should be made to the Publishers.

A CIP catalogue record for this book is available from the British Library.

Shire Library no. 722.   ISBN-13: 978 0 74781 216 6

Kevin Murrell has asserted his right under the Copyright, Designs and Patents Act, 1988, to be identified as the author of this book.

Designed by Myriam Bell Design and typeset in Perpetua and Gill Sans.

Printed in China through Worldprint Ltd.

13 14 15 16 17   10 9 8 7 6 5 4 3 2 1

COVER IMAGE
One of Commodore's later machines, the Amiga 500 – its advanced sound and graphics made it a popular home computer for gaming.

TITLE PAGE IMAGE
Apple's early advertisements for their computers stressed them as general household appliances at home in the living room and kitchen.

CONTENTS PAGE IMAGE
The welcome page to a bulletin-board system heralded a new world of online communications and electronic mail, and the start of online communities for home users.

ACKNOWLEDGEMENTS
My thanks to friends and colleagues at The National Museum of Computing for their help and advice, to Steve Parker for his endless proofreading and to Russell Butcher and Tim Newark at Shire.

IMAGE ACKNOWLEDGEMENTS
Advertising Archive, page 36; Apple, pages 13 and 35; Bill Bartram, pages 24 (bottom) and 26 (top); Marion Butcher, page 26 (bottom); Computer History Museum, page 6 (top); Evan-Amos, page 7 (bottom); Factor-h, page 46 (left); Enrico Gramer, page 30 (bottom); IBM, pages 33 (top) and 33 (bottom); Microsoft, page 37; The National Museum of Computing, pages 5 (top), 5 (bottom), 9, 10, 14, 16 (top), 17 (top), 18 (top), 20, 22, 25 (top), 27 (top), 27 (bottom), 28 (top), 28 (bottom), 30 (top), 32, 34 (top), 38, 39, 40, 41, 42 (top), 44 (bottom) and 45 (bottom); Topfoto, page 46 (right); Neiman Marcus Group Services, page 7 (top); Adrian Pingstone, page 21 (top); PongMuseum.com, page 8 (bottom); Radio Shack, page 15; Science Museum Picture Library, pages 24 (top), 29 and 45 (top).

Shire Publications is supporting the Woodland Trust, the UK's leading woodland conservation charity, by funding the dedication of trees.

CONTENTS

INTRODUCTION	4
THE COMING OF THE MICROCHIP	9
ENTREPRENEURS, ENGINEERS AND ENTHUSIASTS	12
PRACTICAL HOME COMPUTERS	22
IBM AND APPLE SET THE STANDARD	32
GAMES, MODEMS AND THE COMPACT DISC	38
CONCLUSION	46
FURTHER READING	47
PLACES TO VISIT	47
INDEX	48

INTRODUCTION

FOR MOST PEOPLE in the late 1960s and early 1970s, home entertainment was limited to television and the wireless. The wireless receiver had been the centrepiece of the living room and, newspapers and newsreels aside, it was the main source of information and entertainment in the home in the 1950s. Television began to supplant wireless in the 1960s; although it had been invented some time earlier, regular colour television transmissions did not begin in the UK until 1967 – some ten years later than in the US. Initially, colour televisions were prohibitively expensive, but as prices fell, they replaced almost all monochrome receivers by the mid-1970s.

Hi-fi systems were increasingly popular in the home, and many enthusiasts built their own – the same enthusiasts who were also building 'ham' radio receivers and wireless sets at home. Bought music was limited to long-playing vinyl records and the cheaper '45'. Some audiophiles were using reel-to-reel tape recorders to record wireless programmes, but it was the arrival of the cheap compact cassette that allowed young people especially to record music from the radio and share it with friends.

If they were considered at all by the average person, computers were thought to be impressive, mysterious, awe-inspiring and frightening in equal measure. Press coverage of the time typically described new computers as 'electronic brains' and they were often depicted as cartoon machines with faces and arms. Despite the best efforts of engineers explaining their inventions, most people knew more about malevolent computers like HAL from Kubrick's film *2001 – A Space Odyssey*, than the real thing.

In the UK there was at least one well-known benevolent computer: ERNIE. Electronic Random Number Indicating Equipment, or ERNIE

Despite the best intentions of engineers and scientists, computers were often portrayed as menacing electronic brains.

INTRODUCTION

This fully functioning mechanical digital computer was sold as an educational toy in the 1960s. The Digicomp I remains popular and many reproductions have been made.

for short, was a special-function computer designed to generate random numbers as part of the government's Premium Bond saving scheme. Each month the computer would produce a random batch of bond numbers and the winning bondholders would be awarded cash prizes. ERNIE became so well known in the popular imagination that winners would write personally to thank him!

Large mainframe computers were being used by government and industry and almost anyone in work would have been aware of computer systems, even if only from computer printed listings. The machines themselves were carefully watched over by teams of operators, and access to – or even sight of – the machine was strictly limited. For most people outside the computer industry their only interaction would be a badly printed bill from a telephone or power company and the inevitable press coverage of a program bug generating a telephone bill for several million pounds!

The general public began to be concerned about 'number-crunching electronic brains' taking over the jobs of individuals, particularly office workers, but at a time of general high employment, the prospect of increased leisure time outweighed most people's worries.

Although a lot of imagination was required, the Open University BobCat demonstrated the fundamentals of digital computing well before home computers were generally available.

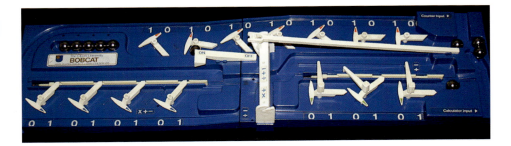

5

The PDP8: the first mass-produced mini-computer from the time of the mini-skirt; it would just fit in the back seat of a Volkswagen Beetle!

Personal computers, that is machines designed to be used by individuals rather than many users at once, were built from the mid-1960s onwards and a typical example is the DEC PDP8 computer. This was a desktop computer designed to be programmed and used by one person at a time. Despite being personal, computers like this were never designed for the home and in 1977 DEC's founder and CEO Ken Olson famously said, 'There is no reason for any individual to have a computer in his home'. In fairness, Olson was probably referring to the possibility of home automation by computer.

Although simple games had been written for almost all of the large mainframe computers, the first recognisably modern computer game was called 'SpaceWar!' which ran in 1962 on a DEC PDP1 mini-computer at Massachusetts Institute of Technology in Boston. This was a two-player game with each player controlling a spaceship displayed on a large circular screen and attempting to shoot each other. Either player could jump

Right: Playing 'Spacewar!' on a PDP1 in 1962 – one of the earliest computer games.

Far right: In this game, two armed spaceships are being controlled by the two players.

INTRODUCTION

to hyperspace and return on a different part of the screen.

One computer advertised specifically for the home during this period was the spectacular 'Honeywell kitchen computer', one of several extravagant gift suggestions in the Neiman-Marcus catalogue of 1969. This was a proper Honeywell 316 minicomputer built into an attractive pedestal unit with a built-in chopping board in front of the control panel. It was advertised for $10,000, but that did include some saved recipes and a two-week programming course! Quite how the home chef would have managed reading recipes through the binary light display is not explained. Sadly, it is not thought that any were ever sold.

Home video-game consoles became generally available in the early 1970s with the launch of the Magnavox Odyssey – a device that plugged into the home television and played a simple bat and ball tennis-style game. The original machines lacked sound and simulated a colour display by including colour transparencies to fix to the front of the television screen!

For the family that had everything in 1969 – the kitchen computer with built-in chopping board.

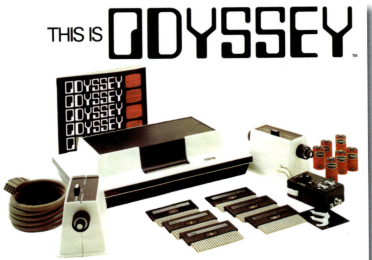

The Magnavox Odyssey was the first home video game.

7

EARLY HOME COMPUTERS

The American engineer, Nolan Bushnell, who had seen the 'Spacewar!' computer game at university and had previous experience with arcade games, founded a new company called Atari in 1972. Nolan and his colleagues produced a coin-operated video arcade game called 'Pong' and had installed their first machine in a bar in 1972. Such was the success of this video game that a modification had to be made quickly to handle the huge number of coins fed into the machine! In 1975, Atari launched a home version of the game called 'Home-Pong' with graphics and sound effects, and was to lead the video game console market for many years, producing many popular machines and games.

Top: One of the earliest arcade video games, Pong was developed by Atari in 1972.

Above: Playing Pong at home on the family television set.

THE COMING OF THE MICROCHIP

Since the development of the transistor, it was possible to build computers smaller and more cheaply than ever. While still typically at least as large as a medium-sized refrigerator, and of limited capacity, desktop computers began to be available to university departments and smaller companies in the 1960s.

Modern integrated circuits, or chips, were first demonstrated in 1958 and allowed engineers to put many transistors on a single silicon chip. Each new chip was typically designed with a single purpose in mind – the processor in a pocket calculator or the sound generator in a video console game, for example.

In 1971, the American company Intel was tasked with designing a new chip for a calculator soon to be launched by the Japanese company Busicom. Rather than creating a new dedicated chip for the job, they instead produced a general purpose device with all the elements of a computer on a single chip. This new chip, or micro-processor, could then be programmed to act as the calculator chip that Busicom needed, but could also be programmed to perform any other function needed by new customers. The new micro-processor, and the many that quickly followed, embodied the fundamental principle of modern computers in that they were general purpose devices – blank canvases on which new tasks could be written.

The outlet for men and boys (and it was almost exclusively men and boys in those days) with an interest in technology in the 1960s and 1970s was through the construction of electronic projects at home – often radios, hi-fi systems and other gadgets that could be soldered together in the garden shed or workshop. Subscriptions for one or more of the widely available electronics magazines were typical, with *Popular Electronics* and *Electronics Today International* being popular titles in the US and the UK respectively.

The 6502 micro-processor chip was the basis for many home computers, including the Apple II.

EARLY HOME COMPUTERS

Famous for its home electronics kits, Heathkit produced several computer kits, including this early system for students.

The design for a circuit would be published in one of the magazines and the eager hobbyist would first visit a local radio store to purchase the necessary parts or perhaps use one of the growing mail-order companies that advertised huge lists of components, and begin the exciting and technically skilled job of building the device. Building a working circuit was just part of the job, and his peers would expect to see the circuitry enclosed in a well-designed and constructed case – typically folded aluminium or later moulded plastic – with neatly labelled controls and indicator lights. The craft skills developed through years of radio and hi-fi construction made the assembly of these new computer kits practical for the hobbyist.

The first complete computer kit advertised was the Altair 8800, which appeared on the front cover of the American magazine *Popular Electronics* in January 1975. Similarly, the first issue of the British magazine *Personal Computer World* featured a British kit computer, which it described as 'perfectly at home'. The Nascom 1 computer on the front cover was a Z80-based kit, available for less than £200.

The computers built were of no real practical use other than to demonstrate the fundamentals of computer design and to act as a learning tool for programming. Once the machine was working, the hobbyist needed to learn how to write a computer program to get anything useful out of it. This was something quite new to most hobby engineers and was often an uphill struggle: it was not unusual for the talented engineer to build the machine, debug all the hardware and then to pass it on to a budding

THE COMING OF THE MICROCHIP

The Altair was first advertised in 1975 and was expected to sell only a few hundred kits, but sold thousands in the first month.

programmer. More often than not, the very early machines had front panels of switches to act as the computer keyboard, and lights to act as the computer display, and a great deal of skill and patience was needed to use the machines at all – not least the ability to think in binary!

Having the customer build and debug the computer at home did make it a lot easier for bright young entrepreneurs and talented engineers to establish companies to market the new computer kits, as a workforce to build or test the machines was not required. A computer company could be formed with simply a good design and a stock of components.

Supporting these budding engineers and programmers were new monthly magazines targeted directly at the emerging market of personal computer users. In the US, *Byte* magazine was first published in 1975 and, in the absence of complete machines to review, typically included articles like 'Choosing a micro-processor' and 'Writing your own assembly program'. Some three years later, in February 1978, *Personal Computer World* magazine was launched in the UK and was quickly followed by similar titles. *Personal Computer World* and *Practical Computing* covered a wide range of machines for both home and business use, and soon magazines dedicated to particular home computers were also published.

These early computer magazines contained a wide range of articles including hardware projects to build at home, program listings for games and applications, which often covered several pages of instructions that needed to be typed in laboriously by hand, reviews of new hardware, and advertisements for component suppliers.

THE BOOK-KEEPER.

THE COOK.

THE CHILD-MINDER.

Meet the ultimate home-help.

THE GARDENER.

THE SECRETARY.

THE TEACHER.

The BBC Microcomputer System.
Designed, produced and distributed by Acorn Computers Limited.

Above are just some of the ways you could use a BBC Micro computer.

And we say 'you' advisedly. For, contrary to popular misconceptions, you don't have to be a technical wizard to use a micro – especially a BBC Micro. Nor do you need any complex equipment.

All you need is an ordinary TV set and a cassette player.

Then with a few basic instructions you can run programs like those above.

There is a huge range of these programs available for the BBC Micro covering games, education and business applications as well as those closer to home.

But, of course, the more you get used to the computer and its language, the more you can get out of it.

To help you do just that, you will receive a step by step User Guide which explains the full capabilities of your micro and shows you how to construct useful programs of your own.

You will also receive a free "Welcome" cassette which contains different programs for you to experiment with, ranging from Music and graphics, to games like Kingdom and Bat 'n Ball.

The BBC Micro is at the heart of the BBC's massive Computer Literacy Project; it is also the most popular and successful machine being ordered by British schools, under the current DOI scheme.

So it is the ideal micro to introduce you – and the family – to home computing. (Although if you have children at school you may find they're ahead of you already.)

The BBC Micro costs less than the average video – only £399. It's available from WH Smith Computer Shops, Boots, John Lewis and local Acorn stockists.

However, if you would like to order one with your credit card or if you want the address of your nearest supplier just phone 01-200 0200.

ENTREPRENEURS, ENGINEERS AND ENTHUSIASTS

IN 1975, TWO CALIFORNIAN SCHOOL-FRIENDS, Steve Jobs and Steve Wozniak, were regular attendees of the Homebrew Computer Club in Stanford, California. The club was a regular meeting of enthusiastic engineers and hobbyists who swapped parts and ideas about computer devices in general. Both had grown up in this area just south of San Francisco that was still to become known as Silicon Valley, but even then it was home to many electronics companies.

In early 1976, the pair began working from Jobs' parents' garage to produce their first computer, the Apple I, with Wozniak assembling the main board of the computer. Neither Jobs nor Wozniak had much previous commercial experience, and even the term '30 Days Net' was a mystery to them. The credit they arranged with the component suppliers gave them just enough time to build the first batch of boards, get paid for them, and pay back their creditors.

Jobs and Wozniak formed Apple Computer Inc. in 1976 and very quickly launched their second model, the Apple II computer. This machine had an attractive moulded plastic case, a full-size keyboard, sound, colour graphics and built-in BASIC, and was ready to go. It was also expandable, with eight expansion slots available to use. The Apple II was advertised widely, and aimed directly at potential home users. With suggested uses such as gaming, teaching arithmetic and spelling to the kids, managing the home finances and even controlling the home environment, the advertising copy advised the buyer to 'clear the kitchen table and bring in the colour TV'!

Opposite: Acorn's machine became known to all as simply the 'Beeb'.

Below: The Apple II defined the standard for personal computers for many years.

One of the first 'plug and play' machines, the PET was complete and ready to go when plugged in.

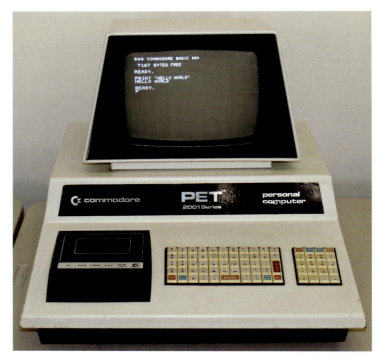

Two other key machines were launched around this time: the Commodore PET, and the Tandy TRS-80. Commodore was an established company well known for the range of electronic calculators it produced, and Radio Shack (known as Tandy in the UK) was a well-known high street retailer of electronics.

Unlike many home-computer manufacturers, Commodore was an established company with a multi-million-dollar turnover selling office equipment and calculators. Jack Tramiel, a survivor of Auschwitz, who had emigrated to the US in 1947, had founded the company in the 1950s. His business growth had been hampered many times by cheaper, better quality products from Japan, and, on a visit there in the 1970s, he saw the first electronic calculators and realised the future of his business with mechanical calculators was going to be limited. Commodore produced a wide range of high-quality electronic calculators using micro-processors from a Pennsylvania-based company called MOS Technology. It was Chuck Peddle from MOS who suggested to Tramiel that computers were the future. Tramiel had seen the Apple II prototype by this stage and might have bought the design but was unable to agree terms with Steve Jobs, so instead asked Peddle to design a new machine. Commodore launched the new design as the PET or Personal Electronic

ENTREPRENEURS, ENGINEERS AND ENTHUSIASTS

Transactor computer in early 1977 and sold it for $595. Prices of components and memory were falling so rapidly at the time that the smaller 4K model was quickly dropped in favour of a machine with twice the memory.

Radio Shack had been established in Boston, Massachusetts in 1921 as a supplier of 'ham' radio equipment – the 'radio shack' typically being the small wooden hut on a ship with the wireless transmitter and receiver. In 1962, Tandy Corporation bought the company, and although the name Radio Shack continued to be used in the US, the name Tandy was used in the UK. The stores catered for electronics enthusiasts and sold components, kits, hi-fi systems and calculators. By the 1970s, Tandy Corporation had thousands of stores worldwide. A big part of the company's profits came from selling CB radio, and as its popularity waned, the company actively began seeking new products. Don French, a buyer for Radio Shack, had bought an Altair kit earlier and set about designing a new machine. Despite the company being somewhat sceptical, they did bring in expert help to design the prototype and French was allowed to continue with the proviso that the end price of the machine was kept low. By February 1977 he showed the prototype computer to the board. They liked it and the finished machine was available in the stores in late 1977. Although the company only expected to sell some 3,000 units in the first year, they actually took orders for more than 10,000 in the first two months. The final selling price, which had been kept as low as possible by removing features such as lower-case letters, was $599 and included a monochrome display and cassette recorder.

All three of these machines became available in the UK by late 1978. The Apple II sold for £1,150, while the PET and TRS-80 sold for around £695 and £499 respectively. These three computers dominated the US home market for many years. The PET and the Tandy machines certainly lacked the colour graphics of the Apple II, but were more affordable and were sold complete with a keyboard, tape cassette player for loading programs, and a display with limited monochrome graphics.

The cost of these three personal computers put them outside the budget of all but the wealthiest of UK home users, although many were purchased for use in education and commerce. However, two British companies, both based in Cambridge, did lead a growing business in producing personal

The Radio Shack, or Tandy, TRS-80 was sold in huge numbers from their many high-street shops.

EARLY HOME COMPUTERS

Similar to Sinclair's MK14, the Acorn System One had a calculator-style keyboard and display.

computers designed particularly for the British home user: Acorn Computers and Sinclair Research.

Sir Clive Sinclair's company Sinclair Radionics Ltd had been designing and selling electronic test equipment, hi-fi amplifiers and, later, calculators since 1961. In 1976, desperate for funds, almost half the company was sold to a government body, the National Enterprise Board. Sinclair had lost control of his company, and as way of continuing his own ideas away from outside interference, he formed Science of Cambridge. Chris Curry, a long-standing Sinclair employee, ran the new company and in 1977 produced a low cost single board computer kit called the MK14. Although the machine was very limited, it did give buyers an inkling of what was to come.

Curry would eventually leave Science of Cambridge, but not until after he and a Cambridge friend Hermann Hauser had formed a new company, CPU Ltd, to further their own ideas in computer development. In 1979, CPU Ltd produced their first computer kit that was sold under their trading name of Acorn Computers Ltd. The Acorn System 75

Above left: Sinclair's pioneering ZX80 home computer was available in the UK for less than £100.

Left: The ZX81, its hugely successful follow-up, sold over 1.5 million units.

ENTREPRENEURS, ENGINEERS AND ENTHUSIASTS

– later renamed as the Acorn System One – was a small 6502-based computer with a calculator-style display and simple keypad.

In 1980 Sinclair famously produced a home computer called the ZX80 for just less than £100. This small, neatly designed computer connected to the home television set and used a standard cassette recorder to load and save programs. The ZX80 had a flat membrane keyboard that was complete although tricky to use. Having a built-in high-level language, BASIC, meant that the system was immediately usable by new owners.

Both Science of Cambridge (later known as Sinclair Computers) and Acorn continued to produce new and innovative designs at a much lower cost than anything available from the US, with Acorn launching their Atom computer, and Sinclair the ZX81 computer.

In the early 1980s the BBC announced the Computer Literacy Project. This was to be a television series with supporting material to explain and demonstrate the possibilities of the microcomputer with particular emphasis on home and education. To that end, the BBC published the specification of a machine they would commission to be built that would be used in the series. Initially the BBC held talks with the UK government-funded company Newbury Laboratories Ltd, and it was

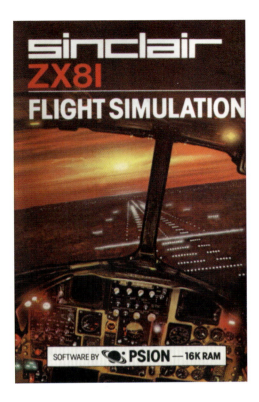

Simulation games have always been popular. This one for the Sinclair ZX81 needs the extra RAM pack.

The job of assembling a home computer from a kit was far from trivial.

17

EARLY HOME COMPUTERS

Newbury Lab's NewBrain computer was the likely contender to be adopted by the BBC.

Acorn's design for a home computer to accompany the BBC Computer Literacy Program was hugely popular in UK schools and homes.

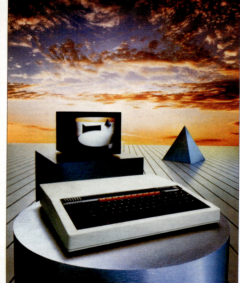

ENTREPRENEURS, ENGINEERS AND ENTHUSIASTS

generally assumed that Newbury would be awarded the contract and their NewBrain computer subsequently adopted by the BBC. However, Newbury pulled out of the possible deal when it became clear they would not be able to produce machines in sufficient quantities. Acorn's Chris Curry was able to persuade the BBC to adapt its specification to allow a wider list of contenders, including designs from Acorn. This opened up the competition to other British companies including Acorn, Sinclair, Dragon and Oric.

Acorn was already working on a successor to their Atom computer to be known as the Proton, and worked hard to adapt the Proton to fit the BBC's specification. Meanwhile, Sinclair worked hard on the ZX Spectrum computer and soon the BBC evaluated both machines. Acorn was successful, and officially awarded the contract with the BBC in 1981.

Despite this setback, Sinclair released the ZX Spectrum with colour graphics as planned, and gradually released upgraded models,

A two-page advertisement for the Sinclair ZX Spectrum.

Sinclair's ZX Spectrum with clearly marked shortcut keys for programming.

19

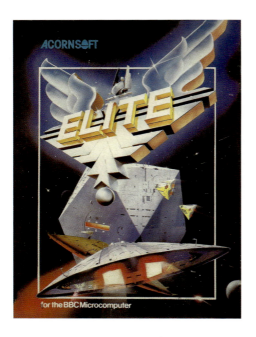 

Above: The seminal space trading game, Elite, was first released for the BBC Micro in 1984.

Above right: A Spectrum cassette game with a tenuous link to a contemporary pop group.

each with more memory, better sound and better graphics. Working with their US partner, Timex, many of these machines were available in the US at very low prices – the Timex branded ZX81 sold many hundreds of thousands during 1982 despite some technical shortcomings.

Almost all of the computers described so far have included the BASIC programming language. This allowed the home user to program the machine, and potentially produce applications of their own. The BASIC computer language had been designed at Dartmouth College in the US in 1964 for a large multi-user system, but the inherently simple and easy-to-learn language specification was made freely available to anyone wishing to implement the language on their hardware. It was rare to write a new BASIC language interpreter, and typically manufacturers bought an off-the-shelf system and had it adapted to their own hardware. The prime purveyor of these BASIC systems was a small software house in Seattle called Micro-Soft. Micro-Soft, later to be renamed Microsoft, was led by a young Bill Gates, who had written a BASIC interpreter in 1975 to run on the Altair computer; this had led to versions that supported more features and many more different machines. Frustrated by the casual attitude to the copying and sharing of his software, Gates wrote an open letter to hobbyists complaining about their behaviour and reminding them of the economics of developing software. Despite this, BASIC remained a core business for Microsoft until the early 1980s.

Acorn's BASIC for the BBC Micro was a radical update to the language in that it included an in-line assembler to take advantage of the machine's facilities and additional language structures, which answered many of the academic criticisms of BASIC. 'BBC Basic', as it became known, has long outlasted the original BBC computer.

The second generation of home computers were sold completed and working, and were of instant practical use. Supplied with a built-in keyboard, the computer could be connected to a standard home television and was good to go, typically running new games, and displaying vivid colour graphics. A colour television solely to be used with the new home computer was often outside the family budget, so the computer was brought in from its original home of the study or garden shed, into the living room, and was connected to the family television set.

These novel and innovative machines had finally become home computers.

Above: Serious concentration was required when playing at home! Note the carefully hand-written labels on the games cassettes.

Left: Joy sticks and light-sensitive 'guns' added to the fun of playing new interactive games.

A Home Finance Package from Acornsoft

PRACTICAL HOME COMPUTERS

RELIABLE HOME COMPUTERS were now available off the shelf, and with a growing library of software and applications they were immediately useful. Nearly all of them supported one or more high-level programming languages and included relatively high-resolution colour graphics. Printers, both 'cheap and cheerful' and high quality, were readily available. Still limiting their usefulness, however, were the slow cassette tape systems used to load programs and to save data.

None of the new companies described previously rested on their laurels but continued to produce new models with better graphics and sound, and an increasing range of off-the-shelf software.

Radio Shack followed up the TRS-80 with enhanced models that were increasingly aimed at the business user. However, their Color Computer or CoCo, launched in 1980, was designed specifically for the home user but had limited success.

The business spreadsheet program 'Visi-Calc' encouraged many businesses to buy an Apple II, which buoyed Apple sales considerably. Visi-Calc is often described as the first 'killer app' – an application that is thought to be so necessary or attractive it is valued more than the core hardware. Apple continued to develop this series of machines until the 1990s.

Commodore quickly followed up the original PET with newer models to answer some of the technical criticisms of the original machine, but made a huge sales breakthrough in 1980 with the launch of the VIC-20. This was an attractively priced machine that was marketed equally in Europe and the US, and was the best-selling home computer in the world with over a million sold by the end of 1982. Commodore advertised the VIC-20 as a more flexible alternative to dedicated games consoles, and continued this argument with the Commodore-64, released at the end of 1982. Named after its 64KB of memory, over seventeen million were sold, making it the best-selling single model computer at the time. In spite of these successes, Jack Tramiel had gradually been losing control of Commodore to his business investors and eventually left in 1984 to form a new start-up company.

Opposite: Among the thousands of games produced for home computers, many productive and useful packages were available.

EARLY HOME COMPUTERS

Commodore's VIC 20 was launched three years after the PET and sold over a million units.

The Commodore-64 and similar machines persuaded potential buyers that an affordable general purpose computer, which was just as good at playing games as running business or education software, was more attractive than buying a dedicated games console. Sales of consoles such as the Atari 2600 and 5200 dropped dramatically between 1983 and 1985, causing many companies to fold or withdraw from the market.

The Commodore 64 was released in 1982 to huge sales throughout the world. It remains the best-selling home computer.

Atari had been producing a successful series of video consoles with increasingly better graphics and sound and with modules that could be added to turn the console into a more general purpose computer, but increasing pressure from successful machines such as the Commodore-64 effectively killed the video console market. Atari's owners at the time, Warner Communications, sold the business to Jack Tramiel who used the sales of the remaining games consoles to fund the development of the Atari ST computer. The ST was one of the first home computers with a colour graphical user interface,

24

PRACTICAL HOME COMPUTERS

nicknamed 'Jacintosh', and a mouse. The nickname was both a reference to Tramiel and to Apple's Macintosh computer. Atari's ST was one of the first home computers to include a MIDI interface which allowed compatible music keyboards and synthesisers to be connected and controlled from the computer.

Just prior to Tramiel taking over, Atari had been funding an ex-employee's start-up company called Amiga Corp. Amiga was developing an advanced chipset for a new range of machines, and Atari was to have use of that new technology in exchange for its support. The change of ownership at Atari prompted Amiga to take the chipset to Commodore instead, rather than share their technology with the new management. When the inevitable legal wrangles were settled, Commodore launched the first of their Amiga series of machines with the Amiga 1000 and continued to develop and sell this range of machines well into the 1990s.

This Atari 2600, produced in 1978, was a popular microprocessor-based video games console and used cartridge-based software.

Atari's first machine with Jack Tramiel at the helm, running their new GUI (graphical user interface) nicknamed 'Jacintosh'.

25

EARLY HOME COMPUTERS

The Amiga 500, launched in 1987, was the first in a series of Amiga machines from Commodore.

Acorn followed their success with the BBC Micro first with a budget version, the Electron, and finally with the BBC Master in 1986. Acorn's later machines, based on their own micro-processor design called the ARM chip, began in 1987 with the Archimedes series of machines and continued until 1992. The ARM chip itself would go on to have phenomenal success in the mobile computing environment.

The Sinclair ZX Spectrum was the first 'Sinclair' model produced after the takeover by Amstrad in 1986.

Sinclair continued to offer upgraded models of the ZX Spectrum series, and launched a brand new machine in 1984 called the Sinclair QL or Quantum Leap. This new design using Motorola's latest high-specification 68000 micro-processor was very ambitious and was launched well before the prototype was finished, let alone available for purchase. Sales failed

PRACTICAL HOME COMPUTERS

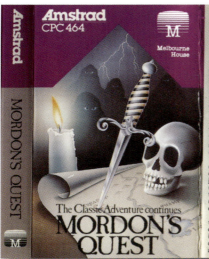

A typical cassette-based adventure game for the Amstrad CPC 464. Despite the macabre cover image, this was a text-based game.

to materialise, partly because of the widely reported problems with the machine, and production was stopped in 1985. Amstrad, which was already a successful consumer electronics company in the UK, acquired all Sinclair's computer products in 1986. Amstrad produced a range of personal computers with their word processor range selling particularly well. They quickly sold off the stocks of older Spectrum and QL machines, and in 1986 launched an updated Spectrum called the 2+. This was followed by further models that continued to sell well in the UK.

Many suppliers developed techniques to make applications such as games and word processors instantly available to the user without the tedious job

ICL took Sinclair's QL computer and added a modem and additional software to make an executive workstation.

27

of spending twenty minutes loading the program from cassette tape. One technique for speeding up this necessary job was embedding the program in read-only memory (ROM) chips. Some manufacturers expected the owner to take their computer apart and plug the ROM chips directly into the computer motherboard, but more often manufacturers supplied the ROM in the form of a plug-in cartridge, which would be connected through a slot in the cover.

Running a word processor by inserting the cartridge in the top of the machine certainly made getting started that much quicker, but did not solve the problems of managing the documents produced – they still needed to be loaded from and saved back to cassette tape.

Supplying programs on plug-in cartridges avoided the tedious job of loading cassette tapes.

The first flexible computer disks, more commonly known as floppy disks, were an 8-inch-square plastic envelope with a flexible circular magnetic disc inside. They were typically included in office computers of the time, but were uncommon in home computers. The smaller 5¼-inch floppy disks, which were cheaper to use, first became available in 1976 but were still relatively uncommon in home computers. Apple successfully brought the cost of the floppy-disk drive down to affordable levels when, in 1978, it developed the Disk II drive for the Apple II.

As the capacity of floppy disks grew, their physical size fell.

Two five-year-old children getting to grips with the first computer in a British infant school.

By the early 1980s, affordable floppy-disk drives were available for most home computers, and their high speed and reliable means of loading new programs and storing data revolutionised the use of the machines – they were now both practical and useful.

The music entertainment industry had for many years given free low-cost records away with magazines. These 'covermounts' as they were known were typically flexible '45s' with new 'pop' recordings. In 1984 the computer magazine publishers followed the same route by taping floppy disks containing games to the front of their magazines. Often these games were copies of the latest retail offerings but with reduced functionality, or games from a software house's back catalogue. Later, compact discs were used instead of floppies, and fully functioning (although time-limited) versions of software were often included.

At the same time as children were beginning to use computers at home, albeit perhaps only to play games, they were also being used in schools. Some schools, particularly those with enthusiastic staff who had previously been in the computer industry, had already installed computers, often previously used machines donated by industry or local colleges. Although the subject of computing was not initially taught as a curriculum item, many school children became proficient programmers through the use of the machines in occasional classes and more often in out-of-hours computer clubs.

EARLY HOME COMPUTERS

Built to stand the rigours of classroom life, the RML 380Z made a home for itself in UK schools.

Many schools had been equipped with slow electro-mechanical teletypes connected across the telephone line to a central mainframe computer perhaps at a local college or university. In many cases this simple slow terminal was enough for schools to run a formal computer education course. Programming was typically in BASIC and despite the very slow speed and the crowds vying to the use the terminal, it introduced a whole generation of school children to programming and computing in general.

By the late 1970s, useful computers for schools were becoming available, and in particular in the UK, the Research Machines 380Z was very popular. This was a well-built, metal-cased machine designed to stand up to the rigours of school use. It included the standard cassette interface for loading programs and included an 8K BASIC interpreter. In 1978 the 380Z was advertised with a single floppy disk drive and 32K of memory for £1,930 in the UK. It was being sold particularly for use in research and education and it was this market that Research Machines continued to service.

The Robotron KC 87 was an East German home computer produced in 1987.

30

PRACTICAL HOME COMPUTERS

East German family life around the home computer was not so different from that in the West: a lot of concentration was required!

An attempt was made, primarily in Japan, to design a standard specification for home computers that manufacturers would adopt. This would allow peripherals such as games controllers and extra memory to be bought from a variety of suppliers. Programs would also be compatible across a wide range of machines. The new standard, called MSX, was championed by Microsoft, although it was little used outside the United States and Europe. Several Japanese manufacturers built systems to the MSX specification, but failed to make worldwide sales.

Throughout this period computer hardware could not be exported from the west to countries in the former Soviet Bloc due to various trade embargoes. Despite this, personal computers were produced there during this period, although the high cost and scarcity of supply meant sales were low compared with the rest of the world. The progression of machines followed a similar pattern to the west, with one of the first models in the Soviet Union, the Micro-80, advertised as a kit in 1983. The Dresden-based East German company, VEB Robotron, produced a wide range of machines including several cartridge and cassette-based personal computers. Limited supplies of these machines meant that most people's contact with the computers was at school or university.

The Elektronika BK series of home computers, produced in the Soviet Union first in 1985, grew to be popular among home users partly as a result of the limited software and support available from the manufacturer. An enthusiastic user community was needed to make useful headway with this home computer, and a cottage industry grew to support the machine.

IBM AND APPLE SET THE STANDARD

B<small>Y THE EARLY</small> 1980s there was a huge range of machines available for the home user, but almost without exception they were mutually incompatible with each other. Not only would programs bought for one machine not work on a different manufacturer's machine; it was also very difficult to share data between systems. Even machines with apparently physically similar disks would not be compatible.

Computer systems designed for small businesses used similar technology to home machines, but most used the Zilog Z80 micro-processor. This standard processor allowed them to use a common operating system called CP/M and to take advantage of the

Left: A typical CP/M-based business computer with twin disk drives built in.

Below: Sharp's MZ80B started with a built-in cassette player, but soon moved towards business users running disk-based CP/M.

wide range of commercial and public domain software produced solely for CP/M systems. More often than not, data could also be exchanged between systems with the use of standard 8-inch floppy disks. Later on, some manufacturers began to extend their machines by using faster 16-bit processors and updated versions of the operating system, which certainly speeded up some ranges, but began to break down the mutual compatibility.

There is almost a myth that IBM either failed to notice the growth in personal computing throughout the 1970s or had simply regarded it as little more than a hobby activity that would not have any effect on mainstream computing or on their own customers. This is far from the truth as IBM had been carefully monitoring exactly what was happening, and in July 1980 began its response with a potential design for a personal computer. For a company that typically took several years to develop a new product, it moved very quickly and IBM's senior managers gave the green light to the proposed design within two weeks; the new computer was in production by mid-1981.

But despite its huge resources, IBM did not have the small, efficient operating system along the lines of CP/M that was needed for the new machine. IBM's software production division was designed to produce software for its large minicomputers and mainframes, not for a small personal computer. The search for an operating system led eventually to Microsoft who did not have a system immediately available, but quickly acquired one which they adapted for IBM's machine and renamed as MS-DOS – which IBM would later rename PC-DOS.

The IBM PC range was launched in August 1981 with the model 5150, typically sold with either one or two floppy disk drives, a small monochrome monitor, a proper office-quality keyboard, and Microsoft's MS-DOS operating system. Quite remarkably for IBM, who were known for carefully protecting their designs and their intellectual property, they made the full specification of the PC available and included full circuit diagrams and listings of its BIOS software. Within weeks of the launch, add-ons to the PC such as extra memory and alternative interfaces were available from other vendors.

Above: IBM's conservative design defined the standards for all PC computers for many years.

Below: A series of adverts was produced for the new IBM PC using Charlie Chaplin's Little Tramp figure.

33

EARLY HOME COMPUTERS

IBM's open design encouraged other companies to produce 'PC-compatible' models, including 'portables' like this early Compaq PC.

Less than a year after IBM's launch, the first PC-compatible computers were released from other companies. These machines were functionally compatible with IBM and could run the same software and use the same add-on components. IBM's choice of format for the already standard 5¼-inch floppy disk led to the disks being forever known as 'IBM format' disks.

Although many retailers tried to sell the first IBM personal computers to home and hobby users, their success was limited, due both to the high price of the machines and their relatively limited capacity when compared with other machines designed for the home market. The format was a huge success in business, however, and dictated the design of desktop business machines for many decades.

Up to this point, few people expected more of their personal computers than being able to type away at a keyboard and see rows of characters appear on the screen. All the interaction with the computer was through typed commands, and of course, the commands varied depending on which system was being used. Graphic displays were available for some computers, but were limited to high-end design and engineering workstations.

Xerox's Alto computer had a GUI display, influencing Apple's design for the Macintosh.

Researchers at Xerox's Palo-Alto Research Centre (PARC) in California had been experimenting with a wholly different approach to interacting with computers by using a graphical representation of documents and files on the computer screen and hand-held devices to point at these iconic representations. Their work had been influenced by research carried out by Douglas Englebart at Stanford Research Institute in the 1960s, and later many engineers had moved from SRI to PARC. In 1973, Xerox launched the Alto personal computer with a high-resolution graphical display, representing the user's desktop, and a pointing device called a 'mouse'. While it was not a home computer, it encapsulated all the elements of what would later dominate our interactions with computers – windows, icons, mouse, and pointing – or WIMP for short.

Apple's designers were planning a new computer, when they were persuaded by Jobs to incorporate some of the revolutionary ideas he had seen in the Alto computer at PARC. The project led eventually to the Apple Lisa computer, launched in January 1983. Lisa was a beautifully designed

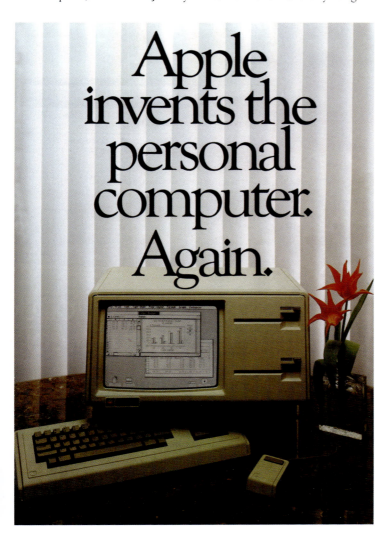

Apple's beautifully designed Lisa computer, which introduced many of the features soon to be seen in the Macintosh.

EARLY HOME COMPUTERS

Apple's Mactintosh was the first computer to introduce the graphical desktop to many users, and proved to be the first in a long series of systems.

and technically very impressive machine, but its high price meant it was never a huge commercial success, and certainly not affordable by the average home user.

While not a simplified or cut-down Lisa in any way, Apple was also working on the design for a new computer to be known as Macintosh; a machine with a graphical user interface or GUI and a mouse that would borrow some of the functionality of the Lisa.

The Macintosh computer was first announced to the media in October 1983, but to the rest of the world by the now famous Ridley Scott television commercial, '1984'. The commercial was transmitted during the US Super Bowl championship on January 22, 1984 and used a figure running towards a huge screen depicting Orwell's Big Brother. The runner launches a missile at the screen as a metaphor of saving humanity from the conformity of the green screen, text-based computer systems of the day.

The screen fades to show the last spoken line in the advert: 'On January 24th, Apple Computer will introduce Macintosh. And you'll see why 1984 won't be like "1984".'

Two days after the television advert was shown the Macintosh went on sale. It included two new applications designed to show off its GUI interface: MacWrite and MacPaint – a word processor and a drawing program. Steve Jobs demonstrated the new computer in the first of his famous Mac Keynote speeches, and although the Mac achieved an immediate and enthusiastic following, it was regarded by many as little more than a toy.

After the initial surge of orders for the new computer, sales rarely achieved the levels that Apple, and in particular Jobs, had hoped for, primarily because of the lack of application software. Since the operating system had been designed primarily to support the new graphical interface, existing text-mode and command-line applications had to be redesigned, and the programs substantially rewritten. This was an expensive and time-consuming task for developers, and was one of the reasons for the initial lack of applications for the new computer. The original Macintosh was also dogged by insufficient memory, and the lack of a second disk drive that made the process of copying disks extremely tedious.

At this stage in history, Microsoft were keen to support the new platform and ported their word processor Word to the Macintosh in the following January; their new spreadsheet program Excel launched first on the Macintosh in September 1985.

Microsoft announced its own GUI interface – Windows 1.0 – to run in conjunction with MS-DOS in 1983, shortly before the launch of the Apple Macintosh, but was not ready to ship the product until late 1985. Windows 1.0 was the first in a continuing series but failed to capture the public imagination until Windows 3.0 was shipped in 1990.

The Macintosh's GUI interface and the availability of creative software for the computer lent itself perfectly to design applications and it became the system of choice for graphic designers and artists for many years. The widely held assumption that Apple systems were only for graphical applications held until the 1990s and polarised the community of users into either Mac users or PC users. Apple would later attempt to dispel this assumption in a series of 'I'm a PC, I'm a Mac' television commercials.

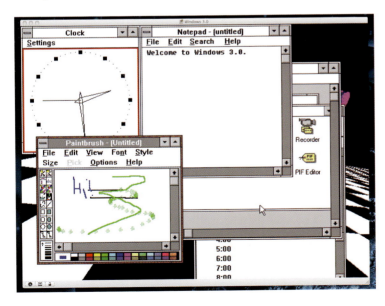

Microsoft's early version of Windows was limited, but by Version 3.0 was becoming useful and popular.

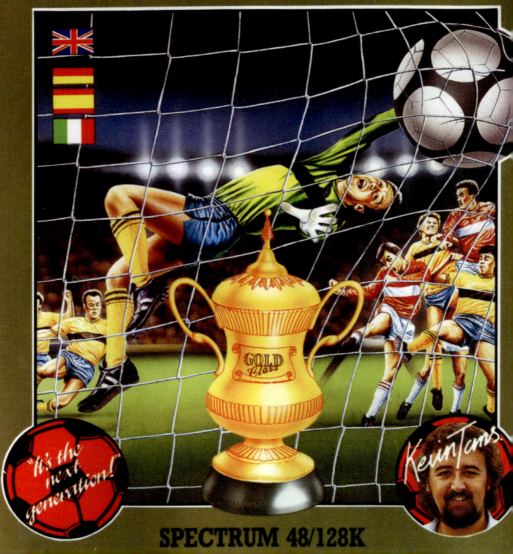

GAMES, MODEMS AND THE COMPACT DISC

Early home computer advertisements always described such worthy applications as storing recipes, managing finances and maintaining the family Christmas card list, but the single most popular use of these early machines was to play games. In the 1970s it was difficult for individuals to write computer games for consoles as they were generally closed systems without the tools needed to produce new software. The advent of practical home computers in the 1980s, with widely published technical details and a wealth of development software, made it possible for individuals to produce and market new software very quickly. Thousands of small software companies were established, many with just one or two employees, and they produced a huge range of software for the home user – primarily games, but also affordable word processing software, music editing programs and simple desktop publishing packages.

Computer games relied on the increasingly better sound and higher-resolution graphics of home computers and many early games were based on those the developers had seen in arcades. For these reasons versions of 'Space Invaders' and 'Pac-Man' were produced for almost all makes of home computer and continued to be popular for many years. Flight simulators had been first produced for the Apple II in 1979, but had become very sophisticated products especially with the launch of Microsoft's Flight Simulator in 1982. MS Flight Simulator would go on to become Microsoft's longest-running software product line.

Adventure games have always been available for home computers, although the early versions were primarily text based. Often provided with very brief

Opposite: Football has always been a popular game for home computers and this game was sold widely in Europe.

Below: Originally written for the Sinclair Spectrum, 'Manic Miner' was ported to many home computers.

user guides, these games would start by leaving the player outside a 'small brick building in a forest' and allow him to move around with simple commands such as 'go north', to solve puzzles and examine and collect objects along the way. 'AdventureLand', produced in 1978, was one of the first text-based adventure games, initially produced for the Radio Shack TRS-80, but soon ported to many other models.

Later adventure games took advantage of higher resolution graphics to allow the player to wander through a virtual world and interact with computer-generated characters. In 'King's Quest', produced in 1984, the player takes the role of the problem-wracked Kingdom of Daventry's bravest knight, Sir Graham. King Edward tells him he has heard of three legendary treasures hidden throughout the land that would end Daventry's troubles. If Sir Graham, in other words the player, succeeds, he will become king!

Developers would continue to battle against software piracy during this time as copying software

Left: 'King's Quest' was a series of graphical adventure games ported to many machines.

Below left: Flight simulators remained popular, and Microsoft's offering was one of their longest-running programs.

Below right: An effective word-processor increased the usefulness of a home or school computer many times over.

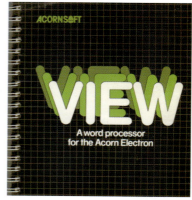

was relatively easy and the user did not regard sharing new games with friends as wrong. Beating any copy protection built into the software was regarded as intellectually challenging and almost part of the game. 'Jet Set Willy', produced for Sinclair's ZX Spectrum in 1984, included a simple effective copy protection scheme. A small card printed with 180 coloured codes was included with the cassette, and on starting the game, one of the codes from the card had to be entered. Although the cassette could be easily duplicated, a copy of the card was also needed, and at the time, colour photocopying was not readily available.

Software designed specifically for schools and home learning was becoming more available during this time and although much of it was designed to support the teaching of computing and mathematics, other subjects were supported. Brøderbund Software's title 'Where in the World is Carmen Sandiego?', first produced in 1985, allowed children to use their knowledge of geography to solve puzzles and eventually find and arrest Carmen Sandiego herself. Popular in UK schools was 'Granny's Garden' (produced for the BBC Micro and later ported to other machines), which allowed the players to progress through a virtual world and find the six missing children of the King and Queen of the Mountains, but only by solving mathematical puzzles and reading and comprehending the text along the way.

In the UK, the availability and simplicity of programming early personal computers, whether at home or in school, enthused an entire generation of would-be programmers and computer professionals. Together with the BBC's Computer Literacy Project, this was a golden time in teaching computing to young people.

Some home computer users had had experience of 'dialling-in' to remote computer services while at school and were keen to use their home machines to access services such as email and news from home. Connecting the home computer to a remote system required a modem, a hardware device for converting digital information from the computer to and from audio tones sent over the phone system.

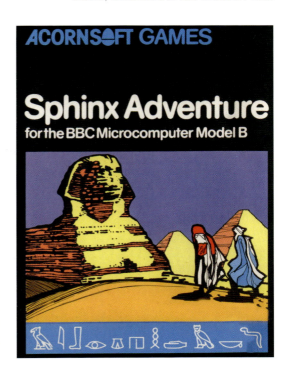

Imaginative design was important in selling these relatively simple text-based games.

EARLY HOME COMPUTERS

A modem was required to connect the home computer to the standard telephone line and communicate with on-line services.

LAST NIGHT WE EXCHANGED LETTERS WITH MOM, THEN HAD A PARTY FOR ELEVEN PEOPLE IN NINE DIFFERENT STATES AND ONLY HAD TO WASH ONE GLASS...

That's CompuServe, The Personal Communications Network For Every Computer Owner

And it doesn't matter what kind of computer you own. You'll use CompuServe's Electronic Mail system (we call it Email℠) to compose, edit and send letters to friends or business associates. The system delivers any number of messages to other users anywhere in North America.

CompuServe's multi-channel CB simulator brings distant friends together and gets new friendships started. You can even use a scrambler if you have a secret you don't want to share. Special interest groups meet regularly to trade information on hardware, software and hobbies from photography to cooking and you can sell, swap and post personal notices on the bulletin board.

There's all this and much more on the CompuServe Information Service. All you need is a computer, a modem,

and CompuServe. CompuServe connects with almost any type or brand of personal computer or terminal and many communicating word processors. To receive an illustrated guide to CompuServe and learn how you can subscribe, contact or call:

CompuServe
Information Service Division, P.O. Box 20212
5000 Arlington Centre Blvd., Columbus, OH 43220
800-848-8990
In Ohio call 614-457-8650

An H&R Block Company

One of the first global dial-up news and email services.

42

GAMES, MODEMS AND THE COMPACT DISC

Although hugely expensive modems had been in use in industry for many years, the first affordable home modems became available in the early 1980s with the Hayes Smartmodem. Hayes developed a standard protocol for communicating with modems that was adopted by very nearly all modem suppliers for the next two decades. Differences in regulations between the US and the UK meant that new products from the US could not officially be used in the UK until approved by UK authorities, although this did not stop many hobbyists from using them in practice.

The early dial-up providers were often commercial computer bureaux that made time available to hobbyists outside their normal business hours. One early system was known as MicroNet, where subscribers could access the system cheaply once they had bought the software needed to connect their home computer. Radio Shack sold a starter kit to allow this, and MicroNet eventually became a US-wide service. In 1987 the service was renamed as CompuServe and soon became available worldwide. Once logged in to the remote system users could send electronic messages between each other, read news, access new software and download patches for their machine. CompuServe would continue until its takeover by America Online or AOL – which would later become infamous for the endless stream of cover magazine compact discs offering their service, many of which later saw good service as bird scarers!

In the UK in 1979, British Telecom launched a commercial online service called Prestel, which included news and email, home shopping and holiday bookings and some premium information services directed at business. In 1983, Prestel also included 'HomeLink', a home banking application for customers of the Nottingham Building Society. This allowed access to statements, bank transfer and bill payments. The relatively high costs of the Prestel system limited its use by home users, and eventually

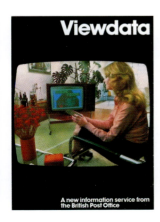

Far left: BT's commercial dial-up information service had limited success with home users.

Left: BT's PRESTEL service was just one of many interactive computer-based information services known generically as Viewdata.

43

Online banking using BT's PRESTEL service arrived in the UK in 1983.

all of these services would become available via free systems.

Hobbyists also used their own home computers to act as the servers in yet more local dial-up systems known as bulletin boards or BBS. Many of these were local, and some very short-lived, but over 100,000 systems were built and advertised widely between early 1970 and the late 1990s. Some of these bulletin-board systems grew rapidly and established connections between each other, allowing the sharing of news and messages and creating chat-rooms for real-time electronic typed conversations between users. Many of the larger BBS systems, such as BIX and CIX, started to offer connectivity to the internet and would later become internet service providers.

New games for the home computer market were being designed to take full advantage of high-resolution graphics and high-quality sound available in all the new home computers. Many games included complete musical soundtracks and even short video clips, but whereas they had previously been distributed on floppy disks, the sheer amount of data needed in these games meant a higher-capacity alternative was needed. The computer industry looked to the audio compact disc that Philips and Sony had introduced in the early 1980s. The compact disc or CD could hold eighty minutes of music, but when used as a digital storage medium, could hold approximately 650MB, equivalent to over 450 standard floppy disks.

As with all new formats for both music and data storage, there exists a 'catch-22' situation where there is insufficient new content available on new media to make it worthwhile for the consumer to upgrade to the new technology, and while there are few consumers with the technology, manufacturers are not keen to invest.

Many adventure games took advantage of CDROMs to include bonus material.

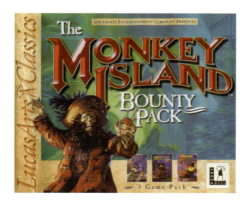

Although early CDROM drives which could be added to home computers were very expensive, two games, 'The 7th Guest' and 'Myst', produced in 1993, were seen as 'killer applications' in that they were so desirable, a drive might well be bought solely for the purpose of playing them. Both were adventure games that allowed the player to wander through a virtual world solving clues and challenges along the way.

GAMES, MODEMS AND THE COMPACT DISC

The high capacity of CDROM drives compared with anything else allowed whole reference books to be delivered to the desktop.

As well as complex games, standard applications and operating systems were also growing in size. The first release of Microsoft's Windows NT shipped on over twenty floppy disks!

Access to large volumes of reference information was not possible for home users before the advent of CDs, other than by accessing an online service that was often slow and cumbersome. A CD version of Grolier's *Academic American Encyclopedia* had been produced in 1985, but did not capture the imagination of the home user, in part because of the high cost of early CD drives. Microsoft released 'Encarta' in 1993 based on a popular American home encyclopedia, but did supplement the content with audio-visual enhancements. *Encyclopædia Britannica*, as printers of the ultimate multi-volume encyclopedia, held out against moving to the new medium to protect the sales of their printed volumes, but finally succumbed in 1996 when a CD version of the encyclopedia was produced and sold for less than $200.

One of the best uses of the new disk format and the sound and video capabilities of the newer home computers was 'Cinemania', released by Microsoft in 1992. This was an interactive database of films and made great use of film clips, stills, excerpts of soundtracks, and complete film reviews.

Games from the pioneering days of home computers still entertain young people, albeit now in museums!

45

CONCLUSION

Below, left: One of the more unusual British home computers was the Jupiter ACE, famous for its use of the FORTH programming language rather than BASIC.

Below, right: A customer tests a Dragon 32 in the shop before buying. Home-computer shops began to spring up around the country very quickly, and some traditional retailers soon began to sell computers.

THE PIONEERING EFFORTS OF ENTREPRENEURS, engineers and programmers in the 1980s took early computer kits and primitive software and developed them into useful and practical appliances for the home. These companies and individuals created machines designed specially for the home user and the vast range of software produced ensured their success in creating this new market. While the home computer market in the US was dominated by a relatively small number of manufacturers, the situation in the UK was quite different, with many companies entering the market, each producing different models with their own strengths and weaknesses. This profusion of designs and their relative affordability enabled any potential owner to buy a computer and to learn to use and enjoy the machine at home.

The introduction of 'industry-standard' designs by IBM and Apple in the mid-1980s would soon dominate the personal computer market and, to a very large extent, remove the distinction between home and business computers, which in turn would signal the end for this imaginative period of home computing.

FURTHER READING

Atkinson P. *Computer* (Objekt series). Reaktion Books Ltd, 2010.
  A thoughtful examination of the physical design of computers.
Barton M. *Dungeons and Desktops: The History of Computer Role-Playing Games.*
  A.K. Peters Ltd, 2008. A detailed examination of adventure and
  role-playing games.
Campbell-Kelly M. and Aspray W. *Computer: A History of the Information
  Machine.* Westview Press, 2004. Very readable history of computing
  with an extensive bibliography.
Donovan T. *Replay: The History of Computer Games.* Yellow Ant, 2010.
  A detailed history of computer gaming, capturing the early excitement.
Laing, G. *Digital Retro: The Evolution and Design of the Personal Computer.* Ilex,
  2004. Detailed photographs of personal computers from the 1980s
  and 1990s.

WEBSITE
The Legacy of the BBC Micro:
www.nesta.org.uk/home1/assets/features/bbc_micro

PLACES TO VISIT

It is advisable to check the opening times before making a journey to visit
  any of these attractions.
The Science Museum, Exhibition Road, South Kensington,
  London SW7 2DD.
  Telephone: 0870 870 4868.
  Website: www.sciencemuseum.org.uk
The National Museum of Computing, Bletchley Park, Sherwood Drive,
  Milton Keynes MK3 6EB.
  Telephone: 01908 374708.
  Website: www.tnmoc.org
National Media Museum, Bradford, West Yorkshire BD1 1NQ.
  Telephone: 0844 856 3797.
  Website: www.nationalmediamuseum.org.uk
Computer History Museum, 1401 N Shoreline Blvd., Mountain View,
  CA 94043, USA.
  Telephone: (650) 810-1010.
  Website: www.computerhistory.org

INDEX

Page numbers in italics refer to illustrations

Acorn 16, 26
   System 1 *16*
   Altair 10
   8800 *10*
Amiga 25
   A500 26
Amstrad *21*
Apple 13, 23, 35
   I 13
   II 13
   Lisa 36
Macintosh 36
Atari 8, 24
   ST *24*
BASIC 13, 20
BBC 17, 41
British Telecom 43
Bulletin boards 43
Cartridges 28
Commodore 14, 23
   C64 *23*
   PET *14*
   VIC-20 *23*
Compuserve 43

CP/M 30
Curry, Chris 16
Digicomp 5
Digital Equipment Company 6
   PDP1 6
   PDP8 6
Dragon *46*
Elektronica 31
Floppy discs 28
Hauser, Hermann 16
Heathkit *10*
Honeywell 7
IBM 33
ICL *27*
Intel 9
Jobs, Steve 13
Jupiter *46*
Magnavox 7
Microsoft 20, 33, 45
MODEM *42*
MS-DOS 33
Nascom 10
Newbury Labs 19
   NewBrain *18*

Peddle, Chuck 14
Pong *8*
Radio Shack/Tandy 15, 23
   TRS80 15
Research Machines 30
Robotron *30*
Sharp *32*
Sinclair 16, 26
   MK 14 16
   ZX80 *16*
   ZX81 *16*
   ZX Spectrum *19*, *26*
Sinclair, Clive 16
Tramiel, Jack 14, 23
Windows 37
Wozniak, Steve 13